RISING TIDE

Designing a Game of International Economic Competition

David R. Frelinger, Jenny Oberholtzer,
Stephen M. Worman, Kate Giglio

Prepared for the Defense Advanced Research Projects Agency

RAND NATIONAL DEFENSE RESEARCH INSTITUTE

For more information on this publication, visit **www.rand.org/t/RRA1947-3**.

Published by the RAND Corporation, Santa Monica, Calif.

Library of Congress Cataloging-in-Publication Data is available for this publication.

ISBN: 978-1-9774-1376-5

Cover: Illustration by Sara Herbst/RAND, Yevhenii/Adobe Stock and royyimzy/Adobe Stock

About This Report

U.S.-China competition, including economic competition, has come to define U.S. foreign policy since 2017. The two economies are the first- and second-largest national economies in the world, and they are deeply intertwined in all aspects of international exchange. Any changes to the relationship, however necessary, could be costly.

To respond to this challenge, RAND researchers conducted economic and institutional analyses of U.S.-China competition, engaged in a participatory foresight exercise to understand the long-term path for ensuring U.S. economic health, and created two economic competition games exploring the dynamics of multiple countries trying to ensure their economic health while interacting with each other and the private sector. The first report in a four-part series includes the economic institutional analyses of U.S.-China economic competition. The second report presents the results of a participatory foresight exercise designed to understand the path for ensuring long-term U.S. economic health. The third and fourth reports describe two economic competition games that explore the dynamics of multiple countries trying to ensure their economic health while interacting with each other and the private sector.

This report, the third in the four-part series, describes *Rising Tide*, an economic competition game, which includes corporations, in which countries pursue their own economic goals without generating conflict with countries that have national goals and political systems that differ within an international economic order that is amenable to change by those countries or corporations. Players acting as countries, regional blocs, or multinational corporations confront multiple challenges related to economic health, including making and sustaining alliances and negotiating competitive globalization, while acting to ensure self-preservation. Participants can opt to play in a system that represents the existing world economic system and work to change the system in small or large ways, or participants can play in an entirely new system to explore what could happen.

Along with the rationale and description of *Rising Tide*, this report includes a general explanation for setup and gameplay and insights drawn from four game test runs. A few simple tips for gameplay are also offered, derived directly from test runs, as well as suggestions for how the game can be developed further.

Although the *Rising Tide* game materials are not available to the general public at this time, we believe that the exploratory experience of designing and testing this game may be of interest to people who are interested in how international dynamics could evolve if different legal, economic, business, and military decisions, policies, and relationships are put into action.

The research reported here was completed in March 2024 and underwent security review with the sponsor and the Defense Office of Prepublication and Security Review before public release.

The views, opinions, and/or findings expressed are those of the authors and should not be interpreted as representing the official views or policies of the U.S. Department of Defense or the U.S. government.

RAND National Security Research Division

This research was sponsored by the Defense Advanced Research Projects Agency (DARPA) and conducted within the Acquisition and Technology Policy Program of the RAND National Security Research Division (NSRD), which operates the National Defense Research Institute (NDRI), a federally funded research and development center sponsored by the Office of the Secretary of Defense, the Joint Staff, the Unified Combatant Commands, the Navy, the Marine Corps, the defense agencies, and the defense intelligence enterprise.

For more information on the RAND Acquisition and Technology Policy Program, see www.rand.org/nsrd/atp or contact the director (contact information is provided on the webpage).

Acknowledgments

We thank our RAND colleagues, Daniel Egel and Bryce Downing, for their respective contributions in developing the Expectations Management Machine and calculator. We also thank Howard Shatz for his project leadership and his willingness to test the game, along with Omar Danaf, William Kim, Gregory Smith, and Karl Mueller. We are grateful to Aaron Frank for his helpful review of a draft of this document. To all involved, well played!

Summary

The United States is no stranger to global strategic competition. For decades, the country has maintained a robust national economy through—and sometimes despite—international trade and political engagement. In recent years, China's economy has been ballooning with profit. Although the Chinese and U.S. economies are highly integrated in many ways, and each country remains the other's most important trading partner, there are challenges in maintaining such a relationship when it impinges on matters of national security. This has led to the emergence of new economic, military, legal, and ethical questions:

- How can the United States *pursue its own economic goals without generating conflict* with countries that have national goals and political systems that are very different than those of the United States?
- Is there a way for the United States to *mutually benefit* from international economic exchanges while maintaining its own interests?

To help economic, business, legal, and military experts explore answers to these questions, RAND game designers developed the tabletop game *Rising Tide*. In *Rising Tide*, participants are designated to play as Brazil, China, India, Japan, Russia, the United States, the Association of Southeast Asian Nations, or the European Union (EU). Participants may also play as one of four multinational corporations. Players interact with one another in a multidimensional, deeply competitive global economic system. They confront multiple challenges related to economic health that pertain to making and sustaining alliances, negotiating competitive deals, building and sustaining alliances, and ensuring self-preservation. Who wins or loses in this game is in the eye of the beholder. The player's goal is to enable their country, regional bloc, or multinational corporation to *survive and prosper, no matter how global tides shift.*

Although the *Rising Tide* game materials are not available to the general public at this time, we believe that the process of designing and testing *Rising Tide* may appeal to people who are interested in how international dynamics could evolve if different legal, economic, business, and military decisions, policies, and relationships are put into action.

Game Design Basics

In *Rising Tide*, as in the real world, national economic health does not exist in a vacuum. Countries and regional blocs have their own political, economic, social, and military interests to pursue, and multinational corporations are always seeking to grow and prosper. Players can pursue any goals via the rule sets that are built into the game, including by wielding soft power or military might over others or by carefully shepherding their country's economy through a major transition period. The game, in its present, first-edition state, does not represent today's world system, although it certainly can. In future editions of the game, designers may create distinctive world systems, complete with increasing modes of digitization, legal and illegal immigration, supply chain factors, and other issues that affect the world. In this iteration, *Rising Tide* allows the game facilitator and, if warranted, the players to decide what the economic world system looks like and then provides a framework to explore how that system might work.

The game includes such details as internal policy choices, sanctions, and non-elite- and elite-population sentiments. These details can affect any economy and thus add real-life nuance. The Expectations Management Machine (EMM)—a spreadsheet based on a simple algorithm—is included to assist country and regional bloc players with decisionmaking and adjudicate player action. The EMM accounts for all various internal and external policy choices a country player has made and helps determine if choices will encourage or reduce economic growth or help a country stay steady. However, the EMM does not provide a strict or precise prediction of how various policies will affect a country. Ultimately, the impetus is on the players to figure out the optimal choices that will drive the best economic outcome for their country because that is one of the goals of the game.

This game was designed from a constructivist point of view that reflects the socially constructed nature of the international system's rules and norms. This view recognizes that the existing international economic system is not a natural state or a given reality. By acknowledging this view from the outset, players can see the system *as one form of many possible forms* and how the system is a product of the countries of which it is composed. In turn, the form of the system shapes the economic opportunities that are available to those countries. The system does not have to exist in its current form, nor is its current form necessarily the best for many of the participants; the system merely holds its current form because of the series of international agreements and actions that led to this moment. In short, *the system does not have to be like it is right now.* Players are given the option to change the existing system during the game or to begin a game that takes place inside an entirely different international economic system than currently exists.

Observations from Game Testing

RAND conducted a full play test of *Rising Tide* in September 2023. The game was played over the course of five hours and consisted of four full game sequences (loops). All game positions were filled, representing eight countries and regional blocs as well as four multinational corporations. After examining play in intervals during and after the game, the following patterns of play emerged:

- **Most players achieved at least one of their initial goal sets by looking after their own well-being.** Rather than focusing on each other or even on the international economic system, these players spent time investing in their allies, feuding with Russia (provoked by a "fangs-out" Russian player, which is not a requirement of the game but was a factor in this round of play-testing), and taking care of internal policies and citizens.
- **It was difficult to entirely transform the international economic system, although this is a game choice.** The option to change the entire system was tried only twice. The first attempt was foiled before the effort even began: The EU player offered incentives to interested parties to leave the system entirely, but no one was interested. The second attempt found the China player facing down the EU and U.S. players, unable to accrue enough power or sway to change things even with several much smaller allies, such as Brazil, backing it up. When the players were asked why there were so few attempts to undermine

or change the international system, they noted that they had cared about their own economies, actions, and goals. Unless changing the economic system was on their to-do list or would help them reach desirable outcomes, players were not interested; there was simply not enough incentive to do so for most players, given the difficulty involved and the opportunity costs.

- **When given the choice to act in a competitive or cooperative manner, the players chose to cooperate.** Some players chose to behave aggressively, but most focused on other options over the course of the test games. The cooperative players chose to build the strongest economies that they could under given circumstances. Competitive activities and punishments were largely focused on responding to behavior deemed actively aggressive. Basic trade disputes were resolved through vigorous discussion and peer pressure rather than attempted overthrows of the international system. The only military actions were the result of early aggression by one country (mimicking recent historical events) and a punishment in response that had been clearly telegraphed beforehand.

Game Play Recommendations

The following recommendations emerged from the observations of game designers during the play test of *Rising Tide*:

- **Create a sizable gaming team with some subject-matter expertise.** The test runs of the game relied on available RAND staff, and only a few participating staff members were senior personnel with deep expertise in the entity they chose to play. This is not to say that the other players were inexperienced or unqualified; rather, a more senior or more specialized group of experts could perhaps have provided more-realistic insights and could have more comfortably explained how and why they made the choices they did. Players with expertise in economics, finance, investment, and international business were especially helpful.
- **Set multiple game objectives rather than just one or no game objectives.** An important early insight from the testing phase was that having pre-set objectives for the players was necessary because the

objectives gave players something to strive for. However, there needs to be balance; pre-set objectives can easily end up as overdetermined win conditions.

The game designers noted that these factors were especially helpful in creating a successful gaming session.

What's Next for *Rising Tide*

The design of *Rising Tide* is flexible, which allows for the exploration of various economic problems and challenges faced by the United States and its partners, both present and potential. As of the time of this writing in 2024, the game team is exploring the possibility of creating a second version of the game. A future version of *Rising Tide* could be crafted to help players uncover the potential impacts of new technologies on and threats to national economies, the international economy, and international relationships. This revision would enable players to delve into such questions as the following:

- How might artificial intelligence, synthetic biology, global environmental change, and other threats affect national systems and the international economic system in the short, medium, and long terms?
- What steps can the U.S. government take now to strengthen its economic and societal well-being and that of its partners in the face of an uncertain future?

By making significant, yet not drastic, adjustments to *Rising Tide* and by incorporating more political and cultural details, the game could evolve into a valuable tool for navigating the complexities of geopolitical uncertainty and substantial change.

Contents

Figures and Table

Figures

Table

CHAPTER 1

Introduction to *Rising Tide*

The United States is no stranger to global strategic competition. For decades, the country has maintained a robust national economy through—and sometimes despite—international trade and political engagement. Since the 1980s, China's economy has been ballooning with economic growth. Although the Chinese and U.S. economies are highly integrated in many ways, and the two countries remain each other's most important trading partners, there are challenges in maintaining the relationship when it impinges on matters of national security.[1] This has led to the emergence of new economic, military, legal, and ethical questions:

- How can the United States *pursue its own economic goals without generating conflict* with countries that have national goals and political systems that are very different than those of the United States?
- Is there a way for the United States to *mutually benefit* from international economic exchanges while maintaining its own interests?

The tabletop game *Rising Tide* was designed to help economic, business, legal, and military experts come together to explore these questions. In the game, specialists with diverse backgrounds can opt to play as a single country, such as the United States or China, as a regional bloc, such as the European Union (EU), or as a multinational corporation. In these roles, the

[1] In a 2023 executive order, President Joseph R. Biden stopped new investments in China's semiconductor, quantum computing, and artificial intelligence (AI) industries on national security grounds (Joseph R. Biden, Jr., *Executive Order on Addressing United States Investments in Certain National Security Technologies and Products in Countries of Concern*, White House, August 9, 2023).

players interact with one another in a multidimensional, deeply competitive global economic system. They confront multiple challenges related to economic health that are related to making and sustaining alliances, negotiating competitive deals, and building up and sustaining alliances while acting to ensure self-preservation. Who wins or loses is not clear-cut because each country's interests are different. The goal of the game is to enable a player's country, regional bloc, or multinational corporation to *survive and prosper, no matter how global tides shift.*

The game was created to help capture the dynamics of human decision-making. Once set up, *Rising Tide* offers players a repeatable experiment that allows them to encounter international economic competition at an active, visceral level. It is like a role-playing game in this way: Players are prompted to embody and thus make decisions for the entity they represent.

Although the *Rising Tide* game materials are not available to the general public at this time, we believe that the process of designing and testing *Rising Tide* may appeal to people who are interested in how international dynamics could evolve if different legal, economic, business, and military decisions, policies, and relationships are put into action. The evolution of this first edition of the game provides experts with a framework to think outside typical academic papers and can aid in the exploration of how international relations might develop within diverse global economic systems.

Pick an Economic World System, Any Economic World System

Rising Tide was designed from a *constructivist* view; that is, it reflects the socially constructed nature of the international system's rules and norms.[2] Any system will do; it is up to the players or the game facilitator to decide which economic world system participants enter, and that system may or may not look like the one that exists. In future editions of the game, the game's designers may create distinctive world systems, complete with

2 See Jack Snyder, "One World, Rival Theories," *Foreign Policy*, October 26, 2009.

increasing modes of digitization, legal and illegal immigration, supply chain factors, and other issues that affect our world.

The game allows players acting as countries, regional blocs, or multinational corporations to follow the rules of the chosen system or attempt to change those rules as they see fit. However, the players do not act alone: Every other player is also pursuing their country, bloc, or corporation's interests. Some of those players' interests may align with a revision or even an overthrow of the current system, while others' interests may be more in line with the established rules because those already suit their needs.

In *Rising Tide*, as in the real world, national economic health does not exist in a vacuum. Countries and regional blocs have their own political, economic, social, and military interests to pursue. Some seek wealth, others seek to wield power, and still others may choose to focus on the well-being of their population or the destruction of their enemies. No matter what is going on, corporations are always seeking to grow and prosper. Players acting as a country, regional bloc, or corporation can pursue any of these goals via the rule sets built into the game by wielding soft power or military might over others in the game or by carefully shepherding their economy through a major transition period.

Nuance and uncertainty are par for the course in *Rising Tide*. The economies of the countries and blocs at play can be affected by the following elements:

- **Policy:** The game's policy cards reflect internal policy choices, such as increased investment in education or infrastructure, a decision to nationalize resources, or a reduction in barriers to trade or immigration.
- **Sanctions:** Sanction cards reflect a trade war and other options for coercive economic measures that are available to players. Sanction actions might target individuals, sectors, multinational corporations, or the economies of countries and regional blocs. The impact is a combination of the level of effort employed and the extent of the trade links.
- **Multinational corporations:** Players can opt to play the game as a multinational corporation. In this role, players can invest in various countries and make deals to ensure a good return on investment. High returns on investment can benefit host-country economies, although that does not necessarily translate to local population benefits.

- **Non-elite and elite sentiments:** Gauging the attitudes of a country's people is one way of scoring in the game. Both non-elite and elite citizens are directly tied to a country's economy but not in the same ways. In the game, an economy that does well will benefit elite populations. We define *elite* as the minority class of people who have the most influence in political and economic decisionmaking.[3] The decisions of the elite do not always have the same effect on non-elite populations as they do on those who make up the elite group. Moreover, certain kinds of noneconomic activity will affect elite and non-elite populations differently. For example, an individual sanction or a targeted nonmilitary hard power action will likely affect elites more directly than non-elites because elites are more likely to be the direct target of those actions. Meanwhile, a direct hard power (military) action will affect non-elites more than elites because elites are more likely to be able to relocate or protect themselves physically than their non-elite compatriots. Not every population in the world will respond identically, so different actions may be necessary to rebalance these scores because these populations might account for different factors.

Although players build on their prior knowledge about the existing economic system in which they must operate (or question), these game elements prompt them to think analytically about the aims and uses of negotiation, soft power, and military might. The Expectations Management Machine (EMM)—a spreadsheet based on a simple algorithm—is included to assist country players with decisionmaking and to adjudicate action. The EMM accounts for all the various internal and external policy choices a country has made. Although the EMM uses a simplified categorization method, it can help determine whether choices will encourage economic growth, reduce economic growth, or help a country stay steady. However, it is not a strict or precise prediction of how various policies will affect a country. The EMM helps, but again, the impetus is on the player to figure out the opti-

[3] A precise and lasting definition of *elite* has been hard to pin down for a long time. For some insight into this difficulty, interested readers should see Paolo Zannoni, "The Concept of Elite," *European Journal of Political Research*, Vol. 6, No. 1, March 1978.

mal choices that will drive the best economic outcome for their country or regional bloc.[4]

Insight into Design: Why a Constructivist Game?

Discussions of competition between the United States and China assume a zero-sum world in which the success of one country necessitates the failure of another. When pundits discuss the possibility of the overthrow of the international economic system, they very rarely underline which elements of this system are to be overthrown. Such discussion, or the lack thereof, rarely suggests that the use of stored value currency will be abandoned or that globalization, industrialization, or free trade are likely to be abolished. The concern is usually more closely tied to the possibility of the United States no longer having dominant control over the system. However, the premise of a system that is dominated by the United States can be challenged; if that premise is true, whether or not such an overthrow would harm the United States could also be challenged.

Rising Tide recognizes that the international economic system is not a natural state. Rather, it is the product of years and even centuries of interaction, extraction, negotiation, exploration, violence, colonialism, internal political upheaval, and technological change. By acknowledging that the system is not sui generis from the outset, players can see that it does not need to be preserved at all costs; rather, they can view the system *as one form of many possible forms* that is the product of the countries and blocs that comprise it. In turn, the form of the system shapes the economic opportunities that are available to those countries and blocs. The system does not have to exist in its current form, nor is its current form necessarily the best for many of the participants; it merely holds its current form because of the series of international agreements and actions that led to this moment.

---

[4] *Rising Tide* is based on an earlier game developed by RAND researchers in 2017. That game was designed to examine a "fangs-out" economic war between rival powers and thus engaged players in mostly deterministic card play. *Rising Tide's* card systems allow for stochastic outcomes and uncertainty.

In short, *the system does not have to be like this*; it is as it is right now, and it may not be this way forever.

However, possible changes to the existing economic system do not constitute a complete destruction of the international economic system. Changes may shift some rules on the outer edge of the system logic structure or, in a more extreme case, reshuffle certain international structures and rule sets. Without hitting a post-scarcity economy, a barter-based system of some dystopic future, or some other option that avoids the need for stored value, international trade, the continued existence of states as political actors, and the basic elements of the current international system seem likely to continue to hold steady. That is, the rule structures and actors do not change just because the economic system changes; merely the numbers, prioritizations, and agents of major changes will shift. Although a rising China may contest the primacy of the United States in this space, it is not doing so to rid the world of international trade or the existence of states.

This explicitly constructivist approach may rankle those who insist on a strict realist understanding of the world, in which a winner necessitates a loser. However, as the name of this game implies, a rising tide can sink many ships, but it can also lift all boats in a harbor. Countries do not need to fight over who will do well in the short term. They just need to try to succeed for themselves.

Rising Tide Allows Players to Explore Malleable International Economic Systems and National Governance

Rising Tide is appropriate for players with different types of expertise who wish to explore, enact, and assess what it means to exist in an international economic system. There is flexibility in the game, presented through different elements cards and option choices. Moreover, players can nominally engage with each other in pretty much any way they see fit: They can cooperate or compete; they can invade, persuade, negotiate, trade, and invest; they can ignore each other and pursue their own ends; or they can team up to bring down the international economic system.

Who *Rising Tide* Is Designed For

The game rules are designed to be flexible, and thus the game is designed for anyone with an interest in international and national economies. After the initial run, the game team revisited the rules in a way that would allow those rules to be bent to reflect any action attempted by players. Because of this, players coming to the game with diverse backgrounds and assumptions based in legal, economic, business, military, and other areas of expertise should feel at ease while engaging in the game. In addition, players with different assumptions can try out their policy choices with the confidence that those choices can be enacted, even if the results are not necessarily what they initially anticipated.

How This Document Is Organized

This document is arranged in four chapters. Chapter 1 provides an overview of the game's purposes and concepts. Chapter 2 presents an explanation of the international system in which the players are immersed and then describes each component of the game that will be used during game play. Chapter 3 offers a general description of game instructions along with visuals, so that players can begin to explore what it means to be part of an international economic system. Chapter 4 presents the findings of the RAND team from the play test of the game. This final chapter also records the changes players made or wanted to make during the game. These are documented for the consideration of gaming design teams, should they wish to refine *Rising Tide.*

CHAPTER 2

Game Elements of *Rising Tide*

This chapter describes each of the *Rising Tide* elements, from the countries and blocs that players can choose to represent, to the factors on which they will base their decisions and the roles that multinational corporations play in the game.

The World of *Rising Tide*

Countries and Regional Blocs in an International System

In *Rising Tide*, every country has its own goals, internal policies to negotiate, and external actions it can apply to other countries. Each country must juggle military costs, sector investments, and non-elite- and elite- population attitudes. This section describes the international system that these countries are working within. We do not focus on a single country; that is the role of the players. Rather, we describe the interactions between countries and the decisions they must make, together and alone.

This game incorporates the world's largest economies at the time of this writing and includes a few smaller, but significant, international players. Alternate player countries were considered during game design, but for the ease of play, not all were kept in the final version. Figure 2.1 shows the countries and regional blocs—the Association of Southeast Asian Nations (ASEAN) and the EU—included in first version of *Rising Tide* as they appear on the "International System Monitor," one of the game's primary boards. Table 2.1 provides brief rationales as to why these countries and regions were included.

FIGURE 2.1
Nations and Regions Included in *Rising Tide*

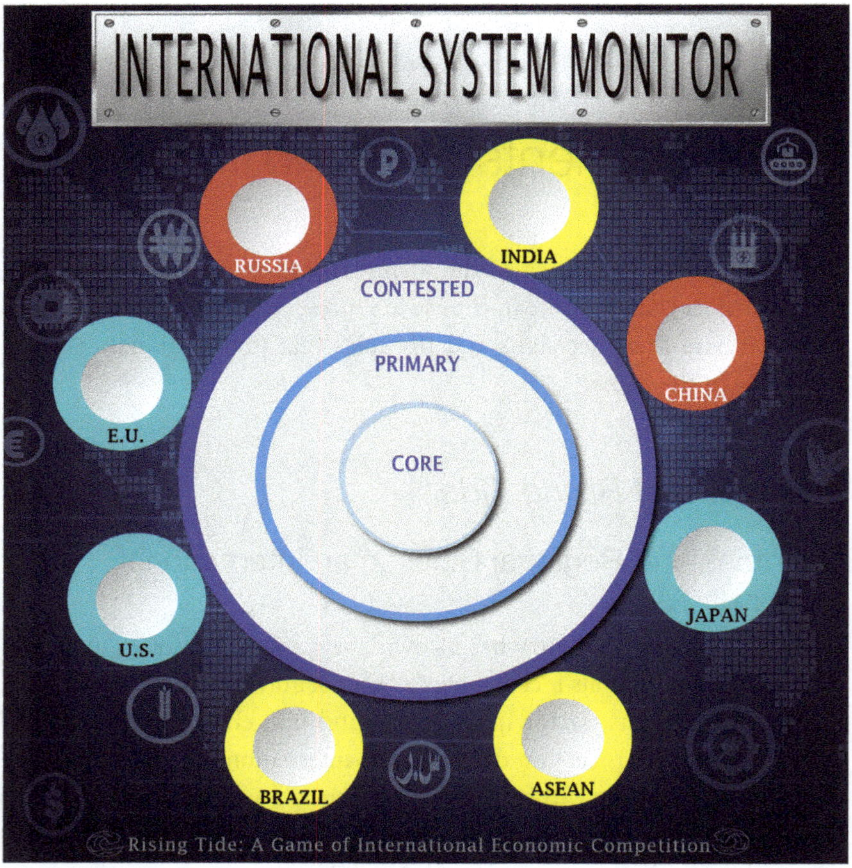

TABLE 2.1
Rationale for Countries Included in *Rising Tide*

Country or Regional Bloc	Rationale for Game Inclusion
EU	Major economies
People's Republic of China	
United States	
Brazil Japan ASEAN	Countries and regional blocs with smaller economies that nonetheless make important decisions that often have international consequences.
India	
Russia	An international spoiler; optional goals can be internally or externally focused, depending on the choice of the player.

NOTE: Regional blocs and countries that were considered but ultimately dismissed included sub-Saharan Africa, South Korea, Latin America (minus Brazil), and Saudi Arabia. The reasons for exclusion were largely size-based: the economic weight of these countries and regional blocs is, in the grand scheme of things, quite small. Other reasons included no clear political or military unity, as in the case of Latin America and sub-Saharan Africa, where convenient geographical groupings does not line up with unified political action. The Middle East as a whole was never considered as a unitary regional bloc because of several major political divisions, limited military power compared with other regional blocs and countries, and comparatively small and nondiversified economies.

As noted in Table 2.1, Russia is included not because it is economically significant but for its role as an international spoiler—a country that can choose to undermine activities pushed by others. India is another example of a country with some economic, military, and social clout; in the game, India's main strength and interest is showing that it can throw its weight around. ASEAN and Brazil are two other economic actors that may have interests in this space but are not yet as large or as strong as India or Russia.

Countries can interact with each other through a series of different actions but also in an informal manner linked to the conversations they have as they negotiate choices. Each player country has their own board to keep track of the different aspects of gameplay. An example of such a board is shown in the bottom half of Figure 2.2.

Country Objective Cards

Each country has a series of optional win conditions on an *objective card*. The top half of Figure 2.2 shows an objective card for the United States. Each objective card offers a few ways to win, as well as a top-level motto about the

FIGURE 2.2
Objective Card Example with Country Game Board

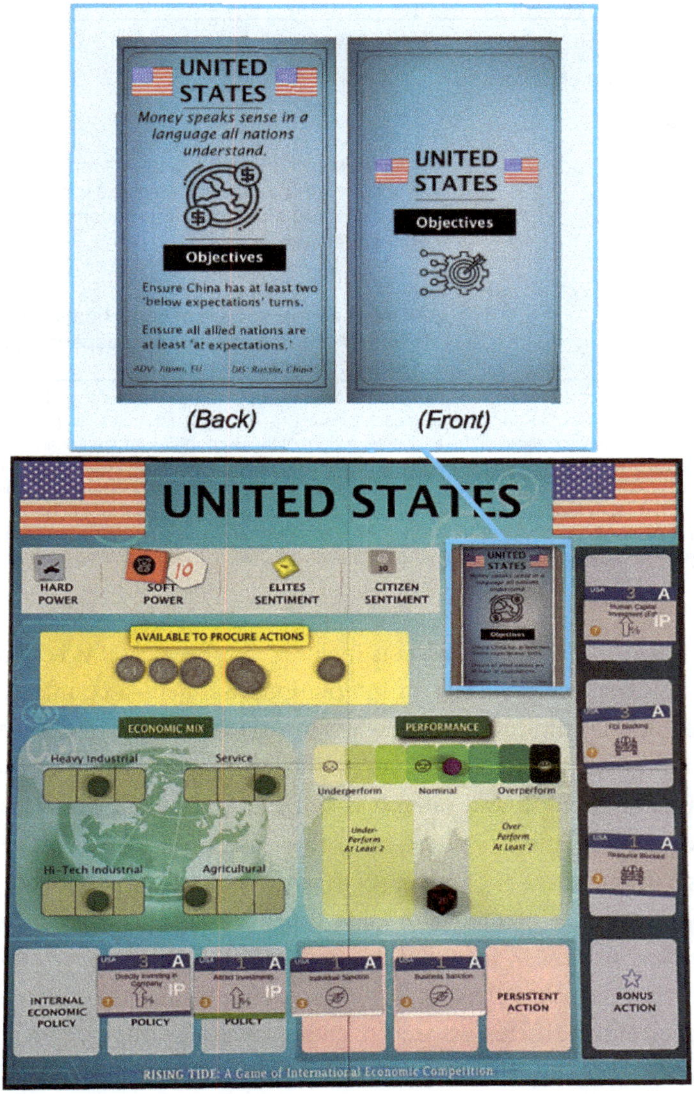

NOTE: ADV = advantage; DIS = disadvantage.

overall game derived from a folk saying in that country. The motto is not vital, but it provides a mental framework for how players can and should approach other players in the game. The objective card also includes information about soft power and relations with other countries. Some countries have a lot of influence over other countries, and some countries have next to none or even negative influence over other countries. Because influence applies to country-country relations at a high level, the objective card is a good location for this information. The contents of the objective card are unlikely to change.

Multinational Corporations in an International System

The *Rising Tide* world is not just made up of countries; it also contains a set of fictitious multinational corporations that act as surrogates for a complex mix of private business entities. These multinational corporations are in fact aggregates of actors with different sets of objectives and thus can form more connections with countries and regional blocs than the countries and blocs on their own. Multinational corporations have different levels of investment and interest in various sectors and countries. Some countries make it difficult for corporations to invest and divest, while others make it easy. The level of ease that corporations experience when they choose to invest or divest in a location can result in a bonus for some countries and a penalty for others. The player countries can set policies that change these bonuses and penalties.

The decision to include multinational corporations with multisector interests, and especially to include *multiple* multinational corporations with different interest mixes, links back to an earlier economic warfare game that featured a single aggregated actor that represented financial and other business interests and was referred to in that game as the business player. The business player could invest in countries and move investments around, but the player did not compete with other business players, nor did the business player have any sway over countries besides their own investment, which would affect actions only through a more-limited set of mechanisms than in *Rising Tide*. By bringing in more corporations that can, for the most part, operate across international borders, we present a bigger opportunity for competition and cooperation and for interesting new dynamics to emerge.

The notional international corporate entities in *Rising Tide* were designed to enhance player engagement; that is, they have been given a different set of general sectors that corresponded to the broad economic sectors that are tracked for the national entities.

Player Interaction

The geopolitical landscape of *Rising Tide* is complex because the world is complex. This complexity is not hidden; in fact, the complexity is front and center because players must negotiate their way through it with their citizens or organizations (if playing as a multinational corporation) and with one another using money, military might, reputation (soft power), and other game elements. Figure 2.3 shows the relationships of the game elements included in *Rising Tide*. Each element is also described in this section.

Players interact with the world and with each other as countries or multinational corporations through the purchase and use of action cards. Action cards can be purchased by players using Business Bucks and Country Coins. Player countries and player corporations keep track of their various scores, policies, and actions on their country or regional bloc boards. The player boards have labeled spaces for statistics, actions, and the Country Coin or Business Bucks they have available. Chapter 3 includes a diagram of a country player board.

Hard Power

Rising Tide addresses economic competition between rival powers and, as in real life, countries sometimes use hard power to interact with one another. In the game, *hard power* is simplified to reflect the expenses involved in increasing military capability and maintaining the same level of military capability. Players can use their military capabilities to attack each other. Much like in the real world, it does not take two players to start a war, but the second party must respond. A simplified military engagement mechanic allows for unusual turns of events, such as small powers holding off large powers, while still portraying the overall dynamic accurately.

A military event can affect hard power (through the loss of military options), soft power (by a player holding their own in the event of an attack), and sentiment (through the targeted use of force on different populations).

FIGURE 2.3
Relationships Between Game Elements

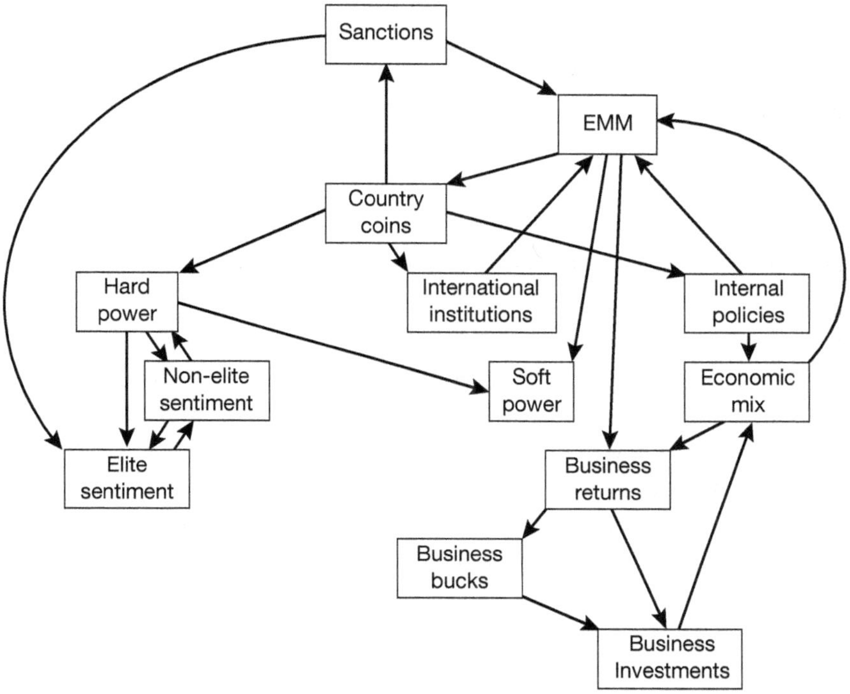

The possibility of devising a plan to annihilate another country does not exist in this game. In its current state, the game includes direct conventional military-to-military clashes, which can, if they turn nuclear, lead to early game termination. Players can attack another country militarily or damage it politically and economically. There is also a nuclear peer escalation rule, under which players can attack another country with their military or damage it politically and economically. Additionally, there is an escalation rule for country and regional powers with significant military potential: engaging in direct combat with each other has a 1/400 chance of triggering an escalation violent enough to end the game as a result of the outbreak of general war. This possibly involves the use of many nuclear weapons for the most powerful countries and regions. However, no one country can outright eliminate another (or a corporation) from the world map. The idea of regime change via conventional military operations was not included in the game design.

Soft Power

In addition to the open negotiation element of the game, countries can encourage each other to reconsider choices. This option reflects informal and back-channel negotiations that allow countries to circumvent the initial choices of leadership. Much like hard power and economic power, not all countries are equal. To reflect the complicated nature of soft power and the history between countries, some countries have a stronger influence over other countries or a lower probability of success depending on rivalries, colonial ties, and historical disagreements. Countries can increase their soft power scores by being good at the actions they take: winning or surviving a hard power attack, persuading others, doing well economically, or even impressing another player enough that they would hand over some soft power points.

Countries can lose soft power by trying any of these things and failing. The stickiness of soft power and the outsized soft power scores of some countries allow for strange elements of the real world to enter the game. Consider the continued cultural influence of several countries that wield high levels of soft power (per a variety of reports and calculations) despite their relatively small size: The United Kingdom, France, Italy, and South Korea all wield levels of soft power that are far greater than their conventional military power or economic might. Military and economic strength alone cannot account for the cultural influence they wield, either historically or recently, and that attracts attention.

Ultimately, *soft power* means being able to unsubtly force a country to make a slightly different decision than it otherwise would have.

The Expectations Management Machine

The EMM is based on a modified Cobb-Douglas formula (a particular equation from economics that describes the relationship between inputs and outputs) and was developed by an economist and a political scientist.[1]

[1] The *Cobb-Douglas formula* illustrates a link between human and physical capital investments on the one hand and economic outputs on the other. See Charles W. Cobb

The EMM exists in spreadsheet form but can be developed into a user-friendly tool. The function of the EMM is to show how various kinds of policy choices will affect a country's economic growth, either positively or negatively, and split that effect into the impact on human and physical capital.

Inputs into the EMM include human capital investment, such as investing in education or attracting migrants; typically, that is anything that involves an investment in the people who reside in a country. Other actions, such as direct investment into a multinational corporation or investment in infrastructure, have similar effects but are classified as influencing physical capital. Although the EMM produces a number representing actual growth (in gameplay terms) relative to expected growth, treating the results of these calculations as accurate predictors of policy outcomes can be problematic because the basis of the model is a highly stylized and abstracted representation of system level effects.

To account for this and avoid inaccurate point estimates, we simplified and narrowed the potential growth outcomes to the following three: likely above, likely below, and likely at expectations. The EMM inputs are the policy and sanctions cards played in the game, from which the effect of the card depends on the size of the economy behind it (the EU throwing its economic weight around versus Brazil), the level of effort (e.g., a comprehensive sanctions regime versus a limited one or a small versus large infrastructure investment), and then the main areas of impact (human capital versus physical capital).

The EMM calculates the cumulative and interactive effects of these estimated impacts and then determines whether the combination of effects is likely to produce economic growth rates above, below, or at the same level as not doing any of those things.

This growth in material well-being affects such things as soft power (a strong performance can impress neighbors) and investment rates from a multinational corporation (it is easier to make an argument to invest in something that is clearly growing well).

---

and Paul H. Douglas, "A Theory of Production," *American Economic Review*, Vol. 18, No. 1, March 1928.

Corporate Investments

Multinational corporations can choose to invest in a country at any given time. Individual countries offer different levels of bonuses to investors but not every sector offers the same return on investment. In addition, corporations that do not have a portfolio in a particular sector will not be able to get a strong return on investment in that sector: A massive return on an investment of zero is still zero. Choosing to invest in a country that aligns with a player's multinational corporation portfolio will have better results.

Lobbying

Although multinational corporations cannot typically make government policies, they can, nonetheless, influence a government's choices. Lobbying cards are a way for corporations to encourage, incentivize, threaten, or strong-arm governments into making policies. Paying to lobby for a particular policy does not mean that the policy will be created, but it does mean that there is a chance that it might be created.

Sectors and Economic Mix

Not all economies have the same mixture of sectors, and not all sectors have the same levels of mobility and growth. To better represent the broad differences between, for example, an agriculture-based economy and a service-based economy, we mark the level of a country's reliance on different sectors: heavy industry, high-tech industrial and manufacturing, service and knowledge, and agriculture. Every country in play has a mixture of sectors listed that reflects the loose grouping of its economy into those four super sectors. The relative strength of those sectors interacts with the level of sector investment from any given multinational corporation to determine the outcomes of investments.

Elite Sentiment

Elite sentiment can be affected directly by certain kinds of sanctions. It can also be affected by hard power actions, such as targeted black operations. Elite sentiment links to general non-elite sentiment, in that any major dip in elite sentiment can be fixed by penalizing non-elite sentiment instead.

Non-Elite Sentiment

As noted, non-elite sentiment is intimately linked to elite sentiment. It can also be hit directly by hard power; for example, people generally do not like it if you attack their country.

Economic Interactions: Internal and External (Sanctions and Policies)

Internal economic policies and *external economic activities* are the meat and potatoes of this game. Both can be purchased and implemented by governments. Their effects are either internal or external: In the real world, economic effects are usually both, but for the sake of game simplification, identifying one or the other was the easiest way to categorize things. These actions ultimately affect the EMM.

Changing the System

Besides one-on-one economic interactions, countries and regional blocs can band together to attempt to change the rules of the international system. They can also choose to oppose changes. An example of two groups is shown in Chapter 3: one group supporting systemic change and another group opposing change square off against one another by rolling dice. The figure does not show the negotiation that happens when the players discuss changes and choose a side for or against. Some changes are complicated and thus harder to put into action than others. For example, changing certain rules around agricultural investments and regulations are simpler to pull off than replacing the U.S. dollar as the main international investment currency. However, even small changes can be relatively expensive and sometimes difficult to enact.

CHAPTER 3

How *Rising Tide* Is Played

The goal of *Rising Tide* is for a player to enable their country, regional bloc, or corporation to survive and prosper, no matter how global tides shift. There can be multiple winners and losers; each player's main task is to explore how actors behave within and against an international system's economic rules and norms. In this chapter, we offer an overview of the game to describe how *Rising Tide* was set up and played during the play test. Please note that this section is not intended to work as a rulebook. We regularly reference the game elements that we described in Chapter 2 of this report.

How to Play

Countries and Regional Blocs

1. **Set up the International System Monitor game board.** The international economic system is portrayed as a three-layered system: an outer layer of rules; an inner layer of norms and institutions; and a core of dollarization, trade, and movement (see Figure 3.1). Game facilitators choose and place the cards that represent the current state of the international economic system.

FIGURE 3.1
Rising Tide **International System Monitor Game Board**

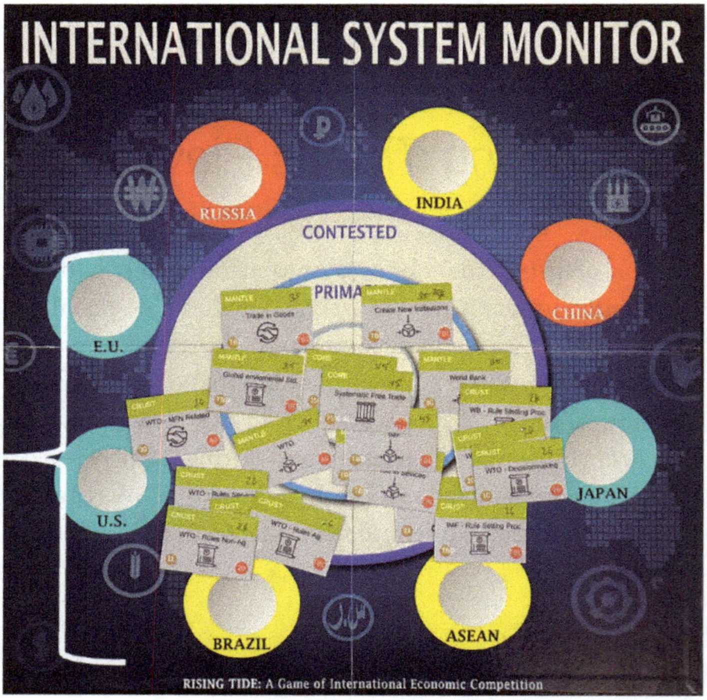

2. **Players or facilitators choose which country or regional bloc the players will represent in the game.** Players can represent one of six different countries or ASEAN or the EU. The participants are ideally selected by the facilitators. However, the assignment of players to a country or a regional bloc should reflect the player's areas of expertise or interest.
3. **Facilitators give each player an initial set of funds based on their country's or regional bloc's economic sector mix and size.** Each country or regional bloc has a different mix of output from one of four different economic sectors: heavy industry, services, high tech, and agriculture (marked yellow in Figure 3.2). This sector mix is predetermined, based on the current state of the world and those countries' economies. This economic mix generates funds, some of

FIGURE 3.2
***Rising Tide* Country Game Board, Card Trackers Highlighted**

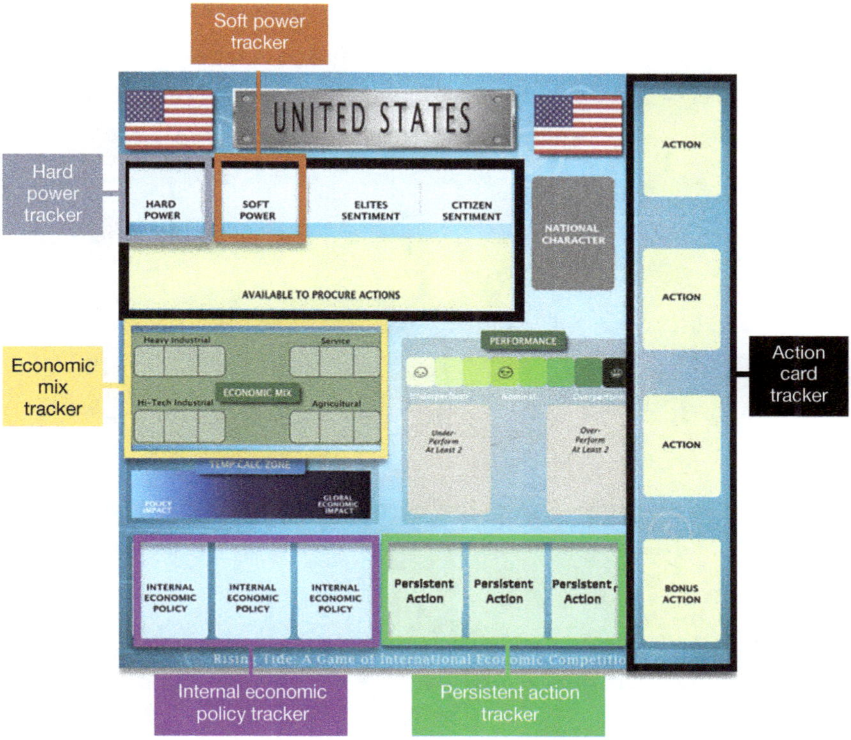

which are siphoned off for hard power upkeep and some of which are used to maintain the country's soft power; the remaining funds are available for player use.

4. **Players begin to use these funds to purchase action cards from their decks of available cards.** Players can purchase as many action cards as they wish from their available cards (see Figure 3.3). However, participants can only play up to three cards at a time under most circumstances, up to the number set by the size of their economy. Purchased cards are placed in the appropriate section of a country or regional bloc board (outlined in black on Figure 3.2).

FIGURE 3.3

Rising Tide Sample Action Cards

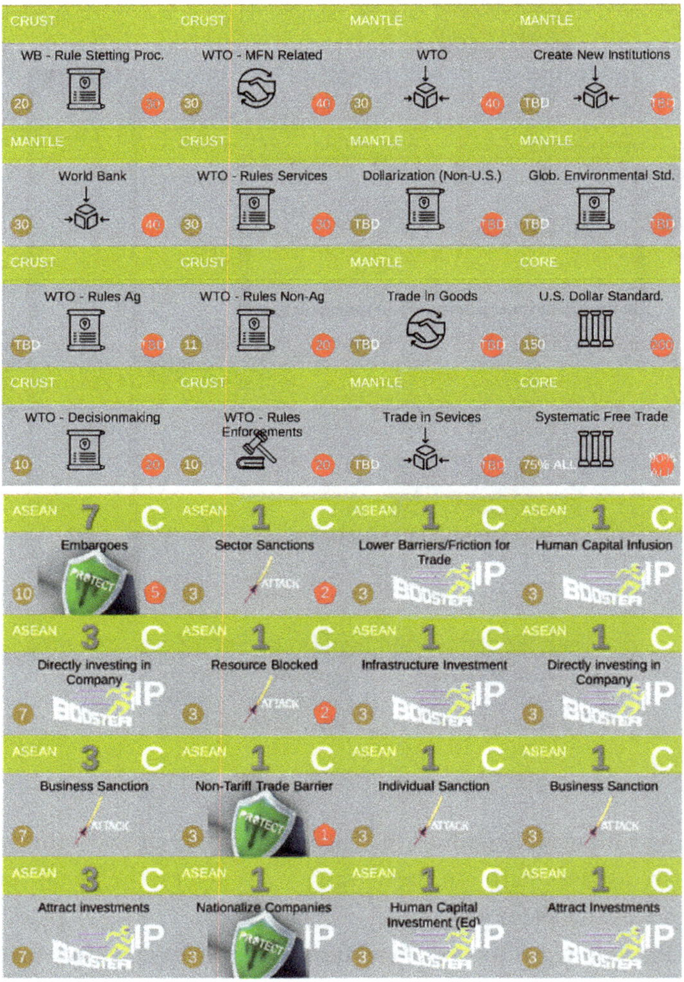

NOTE: Ag. = agriculture; econ = economic; Ed. = education; Glob. = global; MFN = most favored nation; Proc. = procedures; Std. = standards; WTO = World Trade Organization. The geology of a planet is used as a metaphor for the international system in the game. *Core* systems are central: They are exceedingly difficult to change and impact all elements above them. *Mantle* elements are somewhat less important than those in the core system, are easier to change, and only impact the elements above them. *Crust* elements are the least difficult to change but can still affect adjacent elements.

5. **Players negotiate among one another.** At this point, players confer about any actions they plan to take and changes they would like to see. They may also attempt to lure corporations away from competitors. All allowable player actions, except for inter-player agreements, are represented by cards, and thus most negotiations are centered on information offered by a card.
6. **Players take turns playing cards after negotiations.** Cards may be played against or in cooperation with another player, or cards may be played to change the international economic system. Players continue to play cards up to their available card limit.
   a. **Enduring effects:** Up to three cards may be "banked" for an enduring effect on the country. These cards should be placed in the green boxes at the bottom of the player's board (marked green in Figure 3.2).
   b. **Soft power:** Players can force an opposing player to take an action by spending their soft power points (marked in orange in Figure 3.2).
   c. **Hard power:** Countries may spend their hard power points to negatively affect a player in geographic proximity. However, the United States can attack anyone in this game.

7. **Facilitators resolve conflicting card actions.** The facilitator will typically wish to resolve hard power actions first, followed by internal policies and sanctions and then the business cards; soft power and lobbying actions happen at intervals determined by the facilitator. Impacts on the international system and the country are resolved through the use of the country-specific table of effects. The game facilitator determines the results of player actions and calculates the impacts on the economy; the result is whether the country performed according to expectations, over-performed expectations, or under-performed expectations, setting the stage for the next round. Actions against players are resolved directly.

FIGURE 3.4

Changing the System: Order of Play and Simple Decisionmaking Equation

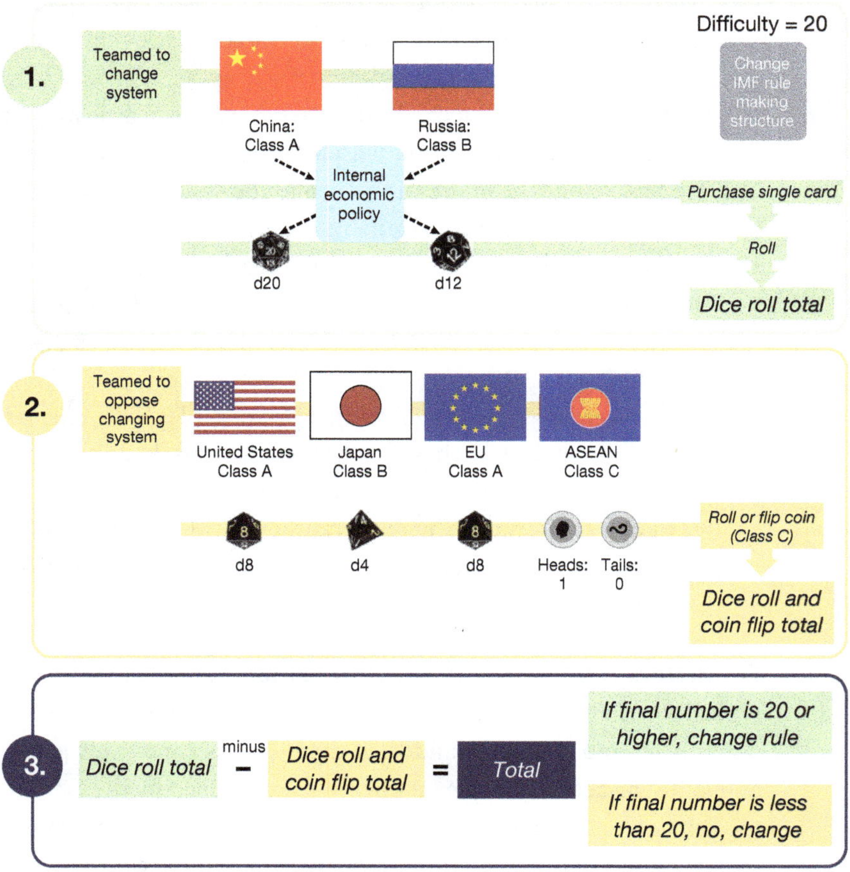

NOTE: IMF = International Monetary Fund.

a. Changes to the international system involve aggregating all cards in favor of change and all cards against change before determining the result. Figure 3.4 charts the way players can work together to promote a change while other players can oppose that change. Dice are rolled by each player country and regional bloc and a simple equation is performed (the third step in Figure 3.4). If the final number is the same or higher as the aggregated changes, then the change can go forward. If the final number is less than the aggregated change number, the proposed change is dismissed.

b. Cards representing rules and norms, which make up an institution, can be changed one by one. Once enough rules are changed, the institution or norm changes. Alternatively, players can attempt to change the norm or institution wholesale with a much lower probability of success.

8. **End of turn: The game facilitator and players make adjustments.** On the basis of player choices, the facilitator marks all changes to the international system. Players adjust the economic output and size of their country's economy as necessary via the EMM and the game loop begins again. The state of the system is recalibrated, and the game continues.

International Corporations

1. **Players who choose to play as multinational corporations get information about their organization.** All information about the corporation and its affiliations with certain countries can be gathered from their corporation player board. Figure 3.5 presents an example of a corporation board.
2. **Players work toward defining and building corporation-country relationships.** Multinational corporations may choose to be located in a country or simply do business with a country. There are benefits and drawbacks to each action.
3. **Players make decisions.** Multinational corporations may decide to relocate offices or extract concessions from countries during negotiations—both actions confer benefits to the host country—and receive profits, given the state of the world.

FIGURE 3.5
***Rising Tide* Sample Multinational Corporation Player Board**

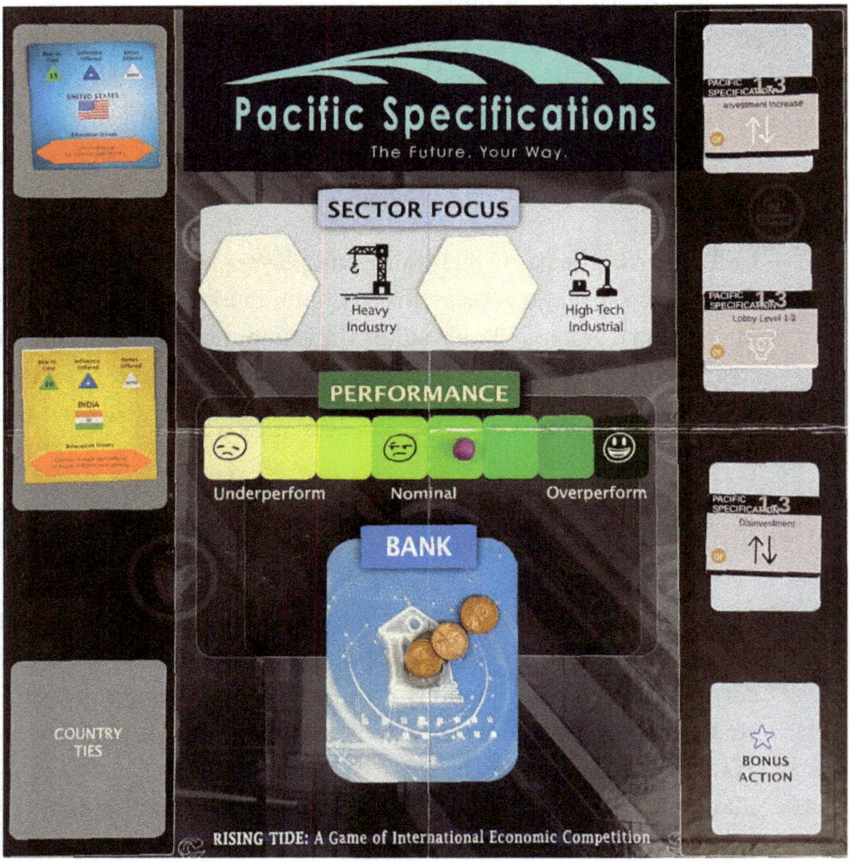

Ending the Game

As with most games designed for concept or scenario exploration, there are no winners or losers in *Rising Tide*. The game ends when the allotted gaming period ends or when players wish the game to end. Game organizers should plan an after-game discussion following the event or as soon as is feasible to capture lessons learned, continue ongoing game discussions, and reflect on the decisions players made and any surprises that occurred during the course of play. If players are operating in teams rather than as individuals, taking notes on team discussions would also be valuable.

CHAPTER 4

Insights from Testing the Game

RAND conducted a full play test of *Rising Tide* in September 2023. The game was played over the course of five hours and consisted of four full game sequences (loops). The players took on all possible positions in the game so that each was filled; that is, individual players took on the roles of the United States, China, the EU, Japan, Russia, India, ASEAN, and Brazil, as well as four multinational corporate entities. Four country and regional bloc players also managed a multinational corporation. Play was conducted with the game designers, who participated as Brazil and ASEAN, two of the smaller entities in the game.

In addition to the full play test, RAND conducted several developmental runs with pairs of players to test dyads and to examine interactions with multinational corporation players. A larger three-country test with one corporate player was also conducted to examine turn speed and ease of adjudication. These developmental tests provided significant insights into how the rules of the game enabled player behavior and helped shape card choices for the full play test.

In this chapter, we first present the outcomes of the *Rising Tide* play test. The intent of this chapter is not to focus on the findings related to gameplay; rather, as the game designers, we sought to ensure that the game dynamics facilitated player collaboration and cooperation in various ways. Should additional runs of the game continue to find the same outcomes, then, in that way, the game is likely very robust. The outcomes related to the international system and the achievement of individual goals discussed in this chapter should not be viewed as conclusive. Those outcomes are recorded to give context to the game dynamics we are interested in. The game was designed to allow players to explore different systems, exchanges, and relationships. There is no doubt that different players will bring varying levels and kinds of expertise and perspectives to the gaming table, which will affect the game's goals and related results in gaming sessions.

After the discussion of outcomes, we then share the considerations and issues that arose throughout gameplay that may inform future renditions.

Game Test Outcomes: Discussion

After several turns, the RAND testing team took a straw poll to see how many players either already had met or were on track to meet their objectives as defined by the win conditions on their objective card. All but two players had already achieved at least one of their initial goal sets. Although some might see this as a reason to adjust the objectives (additional objectives will be added in a later version of the game), it is worth noting that the players who met their objectives represented the United States, the EU, and China. Rather than focusing on one another or even the international economic system, those players spent their time investing in their allies, feuding with the Russia player, and looking after their own well-being.

There were only two instances of players contemplating changing the international economic system. The first attempt was foiled before it even began: The EU player had offered incentives to interested parties to leave the system entirely. The second attempt found the China player facing down the EU and U.S. players but was unable to get enough power or sway to change things, even with several (much smaller) allies backing it up; support from Brazil does not mean much when you are opposed by both the United States and the EU. The goal in this case was not the disestablishment of international institutions or to unseat the U.S. dollar. Instead, it was an attempt to end de facto U.S. veto power in the International Monetary Fund. This attempt took place at the end of the day's gameplay and was less an exercise in setting things up for the next turn and more a matter of curiosity on the part of players. Generally, the players understood that there would be little to no game play impact from the outcome of this attempt, but the exact difficulty level of the shifts in the international system and what affect they would have on the system were matters of interest for the group. The conclusion of the group was that this kind of change is extremely hard to pull off, presumably even without major opposition.

When the question of why there were so few attempts to undermine or change the international system came up, players mentioned their areas of focus: They cared about their own economies, their own actions, and their own goals. Unless changing the economic system was on that list, there was either no hope of doing so (in the case of small countries or a large country that knows it will be opposed by all other large countries) or no reason to do so (in the case of large countries or even medium countries that seem to be doing well). More interesting is that, when choosing to change the international system rather than overthrowing entire institutions or fundamentally breaking rules, these attempts focused on the margins: treating these institutions as legitimate but in need of minor reforms or proposing to change the voting structure or organizing process but never taking down the fundamental institution. This may be because the total elimination of these institutions was set at a higher difficulty level than players were likely to achieve in-game, but players did not disagree with the level setting. In a game in which players were encouraged to offer suggestions and to make rule changes, this kind of agreement either indicates that the players' initial difficulty setting was correct or that there is not enough certainty in this space to enable a more specific answer.

The uncertainty around these costs and probabilities may have been one of the most interesting findings of the internal playthrough. Despite having senior economists and political scientists involved in the development and play-testing of the game, the costs were impossible to agree on because they were so far from the known world. In a fascinating twist, one player managing Russia stated that the list of actions was too limiting: They could not compete with the other countries, given the circumstances. This player had precisely the same options available as all the other countries and had dug a political hole for themselves by engaging in unsuccessful military action early. The political and economic punishment from other players that followed dug the Russia player deeper into the hole, which left them unable to recover for several turns. The Russia player's objectives were only slightly different from the other players' objectives: Their goal was not to grow economically (an objective many countries were charged with) but rather to avoid shrinking economically. The initial move the player made resembled actions by a threatened Russia; in that manner, the player was being a careful student of history.

In the case of the Russia player, the fact that the outcomes caused by a combination of bad luck and retaliatory actions from the actions of other players limited their future options underscores the ways in which player choice can seriously affect game outcomes. The game presents a constructed system based on rules made by a set of game designers, but there is no guaranteed outcome. Had the EU and the United States not decided to punish Russia economically, the option space available to the Russia player would have been far wider. The limitations came from a combination of economic and political factors and the collective actions of other players, who chose to assert their authority and economic strength in the face of undesirable behaviors.

This underlines the importance of human perception in a game like this one. The practical actions that were available at any given time were all the same, but the perception of what was feasible was more important than what was actually feasible. Players tended to focus on what had happened to them recently, what was directly in front of them, and what they could do in response to what had happened in the recent past. The objectives they focused on were not necessarily mutually exclusive. Although those objectives had been selected initially by the design team, the players self-organized in a manner that meant they rarely found themselves at odds while continuing to pursue internal prosperity. In short, the game design allowed the players to act in both a competitive and a cooperative manner, but players chose cooperation more often. Some players chose to behave aggressively, while most focused on their other options and chose to build the strongest economies they could under the circumstances. Competitive activities and punishments were largely focused on responding to behavior that was deemed actively aggressive. Basic trade disputes were resolved through vigorous discussion and peer pressure rather than attempts to overthrow the international system, and the only military actions that occurred were the result of early aggression by one party (mimicking recent historical events) and a punishment in response that had been clearly telegraphed beforehand.

By and large, players focused not on each other but on their own interests and goals, some of which were economic, some of which were focused on prestige or military power, and some of which were quite different. As long as those goals did not come into conflict, there was no reason for the players to cause trouble for each other, and they merely carried on with their moves.

Given the option to compete, cooperate, or carry on, players consistently chose the last two options.

Tips for Playing the First Edition of *Rising Tide*, Drawn from Test Play

There were two conditions for success that game observers noted facilitated the more successful sessions.

1. **Create a sizable gaming team with some subject-matter expertise.** The play test of the game relied on available RAND staff, and only a few participating staff members were senior personnel with deep expertise in the entity they chose to play. This is not to say that the other players were inexperienced or unqualified; rather, a more senior or specialized group of experts could perhaps have provided more-realistic insights and could have more comfortably explained how and why they made the choices they did.
    a. We recommend that game facilitators bring in additional players, especially players with expertise in economics, finance, investment, and international business. In this longer play test at RAND, some players were obliged to double up on roles, running both a state and a multinational corporation. Although this worked well under certain circumstances—the player for China used the associated corporation as an arm of the state, making money for both—for others, it could become a source of confusion or distraction.
2. **Set multiple game objectives rather than one or no game objectives.** The first edition of *Rising Tide* is designed to provide subject-matter experts with a framework to engage as a chosen entity within different international economic systems. Although a particular system is not set in the game—it can be changed to mirror our own historical moment or one in the past or in the future—we found during the testing phase that having pre-set objectives for the players was necessary because they gave players something to strive for. However, there needs to be balance; pre-set objectives can easily end

up as overdetermined win conditions. Without providing multiple options, players can fixate on the single goal provided, even if they are regional experts who should know better. Because set objectives held such outsized power over the players' decisionmaking, the solution was to offer a list of possible goals, which opened up a wider variety of play styles.

Looking Ahead: Possible Uses and Enhancements of the Game

The game design is flexible and allows for the exploration of various economic problems and challenges faced by the United States and its partners, both present and potential. The game team is exploring the possibility of creating a second version of the game. This version would be crafted to help players uncover the potential impacts of new technologies on and threats to national economies, the international economy, and international relationships. This revision would enable players to delve into such questions as the following:

- How might artificial intelligence, synthetic biology, global environmental change, and other threats affect national systems and the international economic system in the short, medium, and long terms?
- What steps can the U.S. government take to strengthen its economic and societal well-being and that of its partners in the face of an uncertain future?

By making significant, yet not drastic, adjustments to *Rising Tide* and by incorporating more political and cultural details, the game could evolve into a valuable tool for navigating the complexities of geopolitical uncertainty and substantial change.

Other changes noted by RAND experts playing *Rising Tide* are being considered for future game versions. Some of these ideas led to the following real-time adjustments during the 2023 play test.

Incorporate specific political and cultural differences into the game. During the play test, some players noted that there was a lack of distinction between authoritarian and nonauthoritarian regimes and a lack of distinct cultural differences that could make game play more accurate or interesting. From economic and military perspectives, these differences may not matter all that much, but they can be a source of frustration and confusion for some players. Although some players with regional- and state-specific knowledge chose actions to best reflect the actions they believed would suit their country's policies, developing country-specific rule sets in later renditions of the game might help improve this capability for all players. For example, the player managing Japan found it surprising when increasing immigration (and thus the foreign population) in a very insular society had no negative political consequences for the government. We could rely on players to represent their interpretation of likely behaviors from the state, or we could provide some country-specific rules that make certain actions more expensive for certain groups.

The existing rule set allows for impact from different soft-power relationships but apart from those rules, there are no major distinctions. In contrast, the earlier game on which *Rising Tide* is based accounted for cultural and political quirks that were specific to countries. For example, that game included a bounceback effect of low non-elite sentiment in Russia, which involved an immediate improvement in sentiment once the lowest possible morale level was hit. The argument supporting this Russia-specific effect is that, historically, this tends to result in the Russian population digging in its heels. Although this rule was initially proposed in jest, several players noted that it accurately reflected historical evidence. Gameplay might be more immersive with several more–country-specific rules that reflect unique or unusual national characteristics. These characteristics are reflected in our country objectives: For example, an unofficial political norm in India is that low morale or even economic instability is fine as long as no famines occur; even though there have been droughts, pandemics, and ethnic cleansings, no famine has occurred in India since the country's independence.

Permit countries to pay for other countries' action cards. Actions considered during game testing included one player investing in another player country's infrastructure or education, but there are certainly other possibilities.

Determine specific rules regarding the ability of multinational corporations to acquire interests in other multinational corporations, to

make sector acquisitions, and more. During game play, it was found that corporate actor rules required more flexibility. Those playing with corporate interests contemplated actions that the game designers and play testers specializing in country actions had not considered.

Enhance player engagement with a single scoring wall that shows the current status of countries' respective economies, sentiment trackers, militaries, and more. This can be accomplished by monitoring the set of boards turn by turn, but players found that approach less desirable. Given that some players had objectives that were tied to the relative performance of other players, a single scoring wall seems like a reasonable approach.

Introduce the creation of free trade associations and other economic or trade organizations as something that players can build. These potential actions might inform longer-term objectives for countries and regional blocs, as well as for multinational corporations.

Increase the sensitivity of elite and non-elite sentiment to force some policy changes. Slow changes in sentiment during game play meant that players could generally ignore them. Increasing the sensitivity of this factor might be especially impactful and accurate when considering elite sentiment. Economic elites' feelings about their own well-being might reasonably link to narrow performance parameters associated with changes in the system, such as financial markets.

Add some specific, large actions to increase soft power. These actions might include space programs, alliance creation or expansion, and other indicators of wealth and success.

Abbreviations

AI	artificial intelligence
ASEAN	Association of Southeast Asian Nations
EMM	Expectations Management Machine
EU	European Union

References

Biden, Joseph R., Jr., *Executive Order on Addressing United States Investments in Certain National Security Technologies and Products in Countries of Concern*, White House, August 9, 2023.

Cobb, Charles W., and Paul H. Douglas, "A Theory of Production," *American Economic Review*, Vol. 18, No. 1, March 1928.

Snyder, Jack, "One World, Rival Theories," *Foreign Policy*, October 26, 2009.

Zannoni, Paolo, "The Concept of Elite," *European Journal of Political Research*, Vol. 6, No. 1, March 1978.

www.ingramcontent.com/pod-product-compliance
Lightning Source LLC
Chambersburg PA
CBHW050520030825
30516CB00012BA/119
* 9 7 8 1 9 7 7 4 1 3 7 6 5 *